SCIENCE FACTORY

CHEMICALS & REACTIONS

JON RICHARDS

PowerKiDS
press.
New York

Published in 2008 by The Rosen
Publishing Group, Inc.
29 East 21st Street, New York, NY
10010

Editor:
Kathy Gemmell

Design:
David West Books

Designer:
Jennifer Skelly

Illustrator:
Ian Moores

Consultant:
Steve Parker

Photographer:
Roger Vlitos

Library of Congress Cataloging-in-
Publication Data

Richards, Jon, 1970–
Chemicals & reactions / Jon
Richards.
p. cm. — (Science factory)
Includes index.
Originally published: Brookfield,
Conn. : Copper Beech Books, 2000.
ISBN-13: 978-1-4042-3906-7 (library
binding)
ISBN-10: 1-4042-3906-5 (library
binding)
1. Chemicals—Juvenile literature.
2. Chemical reactions—Juvenile
literature. I. Title. II. Title: Chemicals
and reactions.
QD501.R5442 2008
540.78—dc22
          2007016564

Manufactured in the
United States of America

INTRODUCTION

This book looks at the basic aspects of everyday chemistry. Different chemical reactions are happening all around us all the time. When ice melts or water boils, a chemical reaction is taking place as the water changes from solid to liquid, or liquid to gas. By following the projects, you can develop your practical skills while at the same time expanding your scientific knowledge.

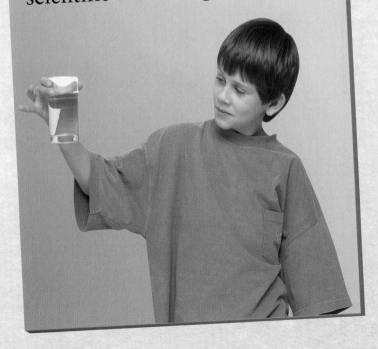

CONTENTS

YOUR FACTORY

BEFORE YOU START any of the projects, it is important that you learn a few simple rules about the care of your science factory.

● Always keep your hands and the work surfaces clean. Dirt can damage results and ruin projects.

● Read the instructions carefully before you start each project.

● Make sure you have all the equipment you need for the project (see checklist opposite).

● If you don't have the right piece of equipment, then improvise. For example, a dishwashing-liquid bottle will often work just as well as a plastic drink bottle.

● Don't be afraid to make mistakes. Just start again — patience is very important!

Equipment checklist:
- Plastic drink bottles, large bowl
- Permanent felt-tip pen
- Salt, fine sand, and soil
- Measuring cup, plastic wrap
- Drinking glasses, tall and small jars, kitchen knife
- Newspaper, paper, blotting paper, cardboard, filter paper
- All-purpose flour, butter, oil, and baking powder
- Dried yeast, fine granulated sugar, lemon, vinegar

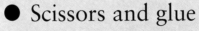

- Red cabbage, fresh milk, baking soda
- Scissors and glue
- Cork and colored cardboard
- Used matchsticks and cotton thread
- Aluminum foil, plastic cup
- Soap and water
- Red food coloring
- Saucepan, long spoon, and colander
- Skewer, splint, feather, and damp dishcloth
- Balloon and candle
- Paint and paintbrush
- Baking pan

WARNING:
Some of the projects in this
book need the help of an adult.
Always ask a grown-up to help you
when you are using scissors or hot saucepans.

FREEZING

WHAT YOU NEED
*Two plastic drink bottles
Permanent felt-tip pen
Water
Measuring cup*

WHEN THE WEATHER GETS REALLY COLD, you may have noticed that the water in puddles, ponds, and even streams turns solid. This solid water is called ice. As the temperature drops, a chemical reaction takes place that changes the structure of the water. We say that the water freezes, or solidifies, to form ice. Look at some of the differences between liquid water and solid ice in this project.

WHY IT WORKS

Water is made up of tiny molecules that float around at random when the water is liquid. When the water freezes, the molecules join into a box-like framework. This keeps the molecules at a greater distance from each other. As a result, water expands when it freezes and occupies a greater volume.

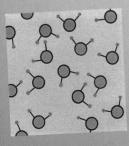

WATER

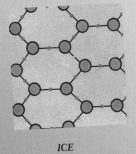

ICE

ICE WATER

Put an ice cube in a glass of water. Ice is made up of molecules that are more spaced-out than the molecules in the liquid water. The solid ice is less dense than the liquid water, and weighs less than the same volume of water. As a result, the ice floats on top of the water.

FEELING THE CHILL

1 Use a measuring cup to pour exactly the same amount of water into each of the plastic bottles.

2 Mark the level of the water on the side of each bottle with a permanent felt-tip pen.

3 Place one bottle in the freezer and leave the other at room temperature in a place that is not too warm. Leave the two bottles overnight.

4 Take the bottle from the freezer and compare the levels of both bottles. You will see that the level in the bottle containing the ice is higher than in the bottle containing the liquid water.

DISSOLVING

WHAT YOU NEED
Three drinking glasses
Water
Measuring cup
Salt
Soil
Fine sand
Plastic cup

IN THE LAST PROJECT, you looked at some of the differences between liquid and solid water. Liquid and solid are known as different states of water. One difference between the two states is that liquid water can absorb and hold certain substances within its structure, but solid ice cannot. This is a chemical reaction called dissolving. However, water cannot absorb every substance, as this project will show.

ONE SUGAR OR TWO?

Look around your home and see if you can find examples of water dissolving other substances. Ask an adult to pour hot water on sugar or instant coffee granules and see what happens.

NOW YOU SEE IT...

1 *Use a measuring cup to pour the same amount of water into each of the drinking glasses.*

2 *Add some salt to one of the glasses.*

3 *Use a plastic cup to pour some soil into the second glass.*

4 Pour some fine sand into the third glass.

5 Stir all three glasses and leave them for a few minutes.

6 Compare the three glasses. You will see that the salt has disappeared. Most of the soil lies on the bottom of the glass. The water with sand in it is cloudy, with a layer of sand at the bottom.

WHY IT WORKS

Salt grains fall apart until they are so small that they can move unseen among the water molecules. They have dissolved. Sand and soil are both made up of small and large particles. Most of the soil particles are too big to dissolve and they sink to the bottom. The fine sand particles are small enough to be carried in the water in a cloudy mixture called a suspension.

SALT

SOIL

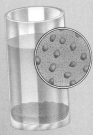

SAND

SEPARATING SUBSTANCES

WHAT YOU NEED
Drinking glasses
Soil
Water
Strainer
Filter paper
Measuring cup

SOME SUBSTANCES DISSOLVE IN WATER better than others. But how easy is it to separate different substances from the water again? The project on this page will show you how to filter a mixture of soil and water. This will separate the mixture to give you what you started with: soil particles and clear water.

WHY IT WORKS

The holes in the strainer are small enough to stop some soil particles from passing through, leaving the mixture a little clearer. The holes in the filter paper are even smaller and only let the tiniest particles pass through with the water. As a result, the mixture is very clear after passing through the filter paper several times.

FILTERING

1 *Pour water into a measuring cup and stir in some soil.*

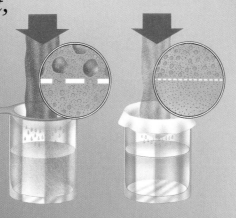

SOIL BLOCKAGE

Repeat the project without the strainer. You may find that the large particles block the holes so that no water can pass through.

2 Place the strainer over a glass and pour the soil and water mixture through it. You will see that the water appears slightly clearer, while the strainer now contains particles of soil.

3 Make a cone out of filter paper. Place it over another glass. Pour what is left of the mixture from the first glass through it. It will become even clearer and you will see tiny soil particles on the filter paper.

4 Repeat with clean filter paper and a clean glass. Be careful not to remove the filter paper too soon, as the mixture may take time to pass through.

5 Repeat the filtering until the water running into the glass is clear. You have now separated the soil from the water.

BOILING APART

WHAT YOU NEED
Drinking glass
Water
Salt
Saucepan

THE LAST PROJECT SHOWED you how to separate soil from water. However, not all substances are as easy to separate from each other. To separate dissolved substances, called solutions, more complicated methods are needed. These methods often make use of differences between the two substances. This project shows you how to separate a solution of salt and water by adding heat.

BRINGING TO A BOIL

1 Dissolve the salt in the water in a drinking glass.

2 Pour the solution into a saucepan. Ask an adult to boil the solution on a stove. It is very important that you ask an adult to do this for you.

3 When all the water has boiled away, ask an adult to turn off the stove and put the saucepan aside to cool.

WHY IT WORKS

As the solution is heated, the tiny molecules in it shake rapidly as the heat gives them more and more energy. The water molecules eventually have enough energy to be able to fly off as a gas called steam. This is called boiling. However, the heat does not supply enough energy for the salt molecules to boil, and they are left behind.

WATER MOLECULES
FLY OFF AS STEAM

4 Look into the saucepan and you will see a layer of white salt. Do not touch it, as it will still be hot.

SUNNY SAUCER

Leave the solution in a saucer on a sunny window ledge. The water will disappear, leaving the salt behind. This works in the same way as boiling the solution, but is slower.

SOAP BOAT

HAVE YOU EVER NOTICED THAT when you fill a glass of water right to the rim, the water appears to bulge above the top of the glass? This is because a force called surface tension holds the tiny molecules of water together, forming a type of skin on the surface of the water. You can break this surface tension by dissolving substances in the water. This project shows you an interesting side effect of this.

SAIL AWAY

1 *Make a boat by pushing four matchsticks into each side of a cork, as shown. Cut out a small piece of the aluminum foil to make a sail.*

WALKING ON WATER

Some insects use the surface tension between the water molecules to actually walk on the water's surface. See if you can spot any of these small creatures skimming across the surface of a nearby pond.

2 *Gently place the boat on one side of the bowl of water. You will find that the boat does not move.*

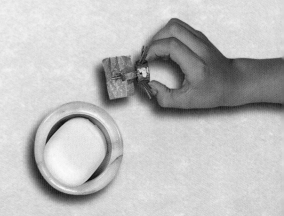

3 *Now dab a small piece of the soap onto the back of your boat.*

WHY IT WORKS

As the soap dissolves in the water, it breaks up the surface tension between the water molecules behind the boat. This means that the water at the front of the boat then has a greater surface tension, which pulls the boat toward it. This causes the boat to move forward.

SOAP

SURFACE TENSION OF WATER

SOAP BREAKS SURFACE TENSION

4 Place the boat in the water again at the edge of the bowl. Watch as the boat moves from one side of the bowl to the other.

WORKING WITH PAPER

WHAT YOU NEED
Newspaper
Flour
Water
Balloon
Oil
Paint
Large bowl
Scissors

MANY MATERIALS THAT WE USE every day are made by mixing and separating natural substances. For example, the paper used to make these pages comes from the fibers of trees that have been separated and then crushed together. This project shows you how combining two substances can create a new substance with different properties.

WHY IT WORKS

When you add flour (1) to water, you get a flour and water paste (2). Putting the paper strips into the paste allows the paste to soak into the pores in the paper (3). As the soaked strips dry, the molecules of water float off, or evaporate (4), leaving a hard substance. This is papier-mâché.

(1) (2)

(3) (4)

PAPER MASK

1 Mix the flour and water in a large bowl to form a thick paste. Blow up the balloon and coat it with a thin layer of oil. Tear the newspaper into strips and soak them in the flour and water paste.

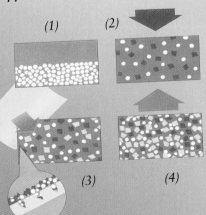

2 Put the soaked strips flat on the balloon. Cover it with about five or six layers of strips. Leave it to dry until the paper has become very hard. This hard substance is called papier-mâché.

3 Ask an adult to carefully cut the papier-mâché from the balloon. The oil should make this a lot easier, but it doesn't matter if the balloon pops. Ask an adult to cut the shape in half to form the mask and then to cut out the features.

PAPIER-MÂCHÉ FUN

Think of other objects to make from papier-mâché. You could make a wastepaper basket, using the basket instead of the balloon as your mold.

4 Decorate your mask with paint. Make sure the paint isn't too wet.

5 If you like, you can turn the hard papier-mâché back into paper and paste. Fill another bowl with water and simply soak the mask in the water.

6 The mask becomes floppy when you soak it in water. You have now reversed the papier-mâché-making process.

ADDING HEAT

WHAT YOU NEED
Flour
Water
Large bowl
Skewer
Cotton thread
Paint

SOME FORM OF ENERGY IS OFTEN needed for a chemical reaction to start. The heat in a room may be enough for a substance or mixture to start changing, but in many cases, more heat is needed. For example, china plates and cups have been baked in a very hot oven called a kiln. This turns them from soft clay into hard china. This project shows how heat can turn flour and water into a hard substance.

COOKING JEWELRY

1 Mix flour and water together in the large bowl to make a dough. Knead the dough well to squeeze out any air bubbles.

2 Shape and roll the dough into beads like these. Sprinkle flour on your work surface to stop the dough from sticking. Ask an adult to make a small hole in each bead with a skewer. This will let you thread the beads when they are baked.

3 With the help of an adult, cook the beads in an oven until they harden. Allow them to cool, then push the thread through them to form a necklace.

4 Decorate the necklace with the paint.

MAKING IT EASIER

Change the dough ingredients by adding a little oil. This makes it easier to mold into shape.

WHY IT WORKS

When flour is mixed with water it forms a paste or a dough, in which the flour molecules are suspended in the liquid. The consistency of the dough depends on how much flour you add. When the dough is baked in the oven, the heat causes the water molecules to evaporate (1), leaving behind a hard substance (2).

(1)

(2)

5 Wear the necklace yourself or give it to someone as a present.

SECRET MESSAGES

WHAT YOU NEED
Lemon
Feather
Sheet of paper
Candle
Jar
Kitchen knife

HEAT PLAYS AN IMPORTANT ROLE in many chemical reactions, not just in turning flour and water into hard clay, as you discovered in the last project. You can see how important heat is every day in any cooked food that you eat. On pages 28-29 you will see how heat is used to make bread. This project shows how heat can turn something that was invisible into something you can see.

INVISIBLE WRITING

1 *Ask an adult to cut a lemon in half. Squeeze the juice out of it into a jar.*

2 *Place the sheet of paper for a little while in a warm oven that has been turned off. This will turn the paper very slightly brown and make it look old.*

3 *Ask an adult to trim the end of the feather to make a quill pen. Dip the pen tip in the lemon juice.*

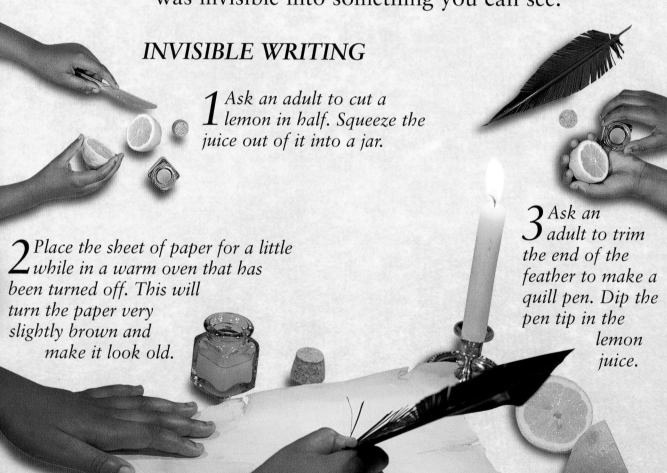

4 *Write your secret message on the browned sheet of paper.*

MAKING TAFFY

Ask an adult to help you use heat to make taffy. Mix sugar, water, and a drop of vinegar in a saucepan. Bring to a boil, stirring all the time. Remove it from the heat and add baking soda. Pour it into a buttered pan and leave it to cool.

5 To read your message, simply hold the paper close to a lit candle. Ask an adult to do this for you. The message will appear on the paper.

WHY IT WORKS

The heat from the candle causes water molecules in the lemon juice to evaporate. It also causes the substances left on the paper to react with oxygen in the air. This process, called oxidation, turns the lemon juice brown, making it visible.

OXIDIZED LEMON JUICE

WATER MOLECULES EVAPORATE

ACIDS AND ALKALIS

ONE WAY OF DESCRIBING CHEMICALS is to say whether they are acids or alkalis. Acids, such as lemon juice, tend to have a sour taste, while alkalis, such as milk, tend to be slightly soapy to the touch. Pure water is neither acid nor alkali — it is described as neutral. Don't try tasting or touching chemicals at home, though. This project shows you a simpler and safer way to test whether a liquid is an acid or an alkali.

WHAT YOU NEED
Red cabbage
Kitchen knife
Saucepan, Jars
Water, Lemon
Blotting paper
Vinegar, Milk
Baking soda
Paintbrush

FOOD INDICATORS

You can make indicators from other fruits and vegetables. Blackberries and blueberries make good indicators and so do beets. Use these to test other liquids around the house.

DETECTING ACIDS

1 *Ask an adult to cut the cabbage into small pieces. Put it in the saucepan and cover it with water.*

2 *Ask an adult to heat the saucepan on the stove. Boil the water and let it simmer for 10 minutes. Remove it from the heat and let it stand for one hour.*

3 *Drain off the cabbage juice. Dip the blotting paper into it and leave it to dry. This will be your indicator paper.*

4 *Squeeze the juice from the lemon. Pour it, the milk, the vinegar, and the baking soda into separate jars.*

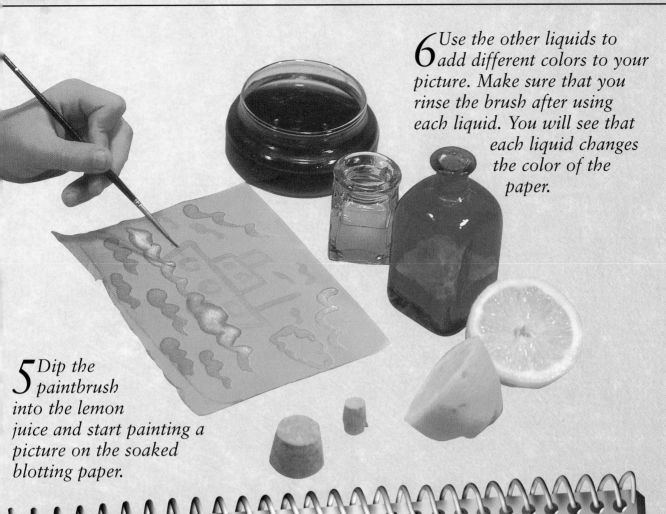

6 Use the other liquids to add different colors to your picture. Make sure that you rinse the brush after using each liquid. You will see that each liquid changes the color of the paper.

5 Dip the paintbrush into the lemon juice and start painting a picture on the soaked blotting paper.

WHY IT WORKS

The juice from the cabbage is a good indicator of how acidic or alkaline a liquid is. The indicator reacts with the liquid and changes color according to whether the liquid is acid or alkali. If the liquid is an acid, the paper will turn red or pink. If the liquid is an alkali, the paper will turn blue or green.

ACIDS ——————— NEUTRAL ——————— ALKALIS ▶

BUBBLING VOLCANO

WHAT YOU NEED
*Colored
cardboard
Plastic cup
Glue, Water
Newspaper
Paint
Baking soda
Vinegar
Red food
coloring*

IN THE LAST PROJECT, you looked at differences between acids such as vinegar and alkalis such as milk. But what would happen if you were to mix acids and alkalis together? Farmers do this regularly, by adding alkaline lime to acidic soil to make the soil neutral. This project shows what happens when you mix acidic vinegar with alkaline baking soda — the effects are spectacular.

CHEMICAL VOLCANO

1 *Glue a plastic cup to a cardboard base. Make a cardboard cone, wrap it around the cup and glue to the base.*

2 *Cut off the top of the cone, then glue newspaper strips onto it. Glue on five or six layers of strips. When the strips are dry, paint your volcano and let it dry.*

3 *Mix three parts water with one part glue and paint the volcano with the mixture. This will protect the cone when the volcano erupts. Let this dry.*

WHY IT WORKS

When the acidic vinegar and the alkaline baking soda mix they cause a reaction that releases bubbles of a gas called carbon dioxide. These bubbles of gas cause the mixture to foam up and erupt out of your volcano.

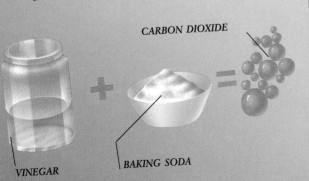

CARBON DIOXIDE

VINEGAR   BAKING SODA

4 Mix some red food coloring with the vinegar. Put a teaspoon of baking soda into the cup in the volcano.

5 Now pour in the vinegar and food coloring mixture.

FIZZY LEMONADE

Bubbles of a gas called carbon dioxide are what make fizzy drinks fizzy. To make fizzy lemonade, mix the juice of four lemons with water and some sugar. Stir in a little baking soda. Watch it fizz, and then drink it!

6 Watch as red foam erupts out of the volcano, like red-hot lava from a real volcano. The more chemicals you use, the bigger the eruption will be.

GIVING OFF GAS

WHAT YOU NEED
Cork
Used matchsticks
Small candle
Tall jar
Long spoon
Splint
Baking soda
Vinegar, Water

WHEN A FIRE BURNS, a chemical reaction takes place between the burning substance and a gas in the air called oxygen. If there were no oxygen, the fire could not burn. To put out fires, firefighters spray water, foam, or even carbon dioxide gas to prevent more oxygen from reaching the fire. This project shows how carbon dioxide can extinguish (put out) a lit candle.

FIRE EXTINGUISHER

1 *Make a small boat using the cork and used matchsticks as shown on pages 14–15. Stick on the candle instead of the mast and sail. Float your boat in some water at the bottom of a tall jar. Ask an adult to light the candle using a splint.*

DIFFERENT SHAPES

Think about which places might have fire extinguishers. Look around at the different kinds. How many shapes can you can find? Never touch a fire extinguisher, as it might go off.

2 *Carefully add several spoonfuls of the baking soda to the water. Stir the mixture gently with a long spoon.*

3 Quickly pour some of the vinegar in the jar, making sure that you don't spill any on the lit candle. The liquid should begin to fizz. If it does not, add more baking soda and vinegar.

4 Watch as the liquids at the bottom of the jar fizz and bubble. The candle will dim and finally go out.

WHY IT WORKS

Mixing the vinegar and baking soda together creates bubbles of carbon dioxide. As these bubbles rise from the liquid, they push out the air that was in the jar. Without oxygen in the air, the candle can no longer burn, so it goes out.

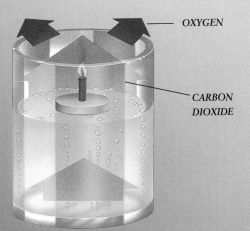

OXYGEN

CARBON DIOXIDE

MAKING BREAD

ALL OF THE REACTIONS YOU HAVE SEEN involve nonliving chemicals. These are called inorganic reactions. There are other types of chemical reactions involving living organisms. These are called organic reactions. This project shows you how bread is made and how the chemical reactions of a tiny mold called yeast make bread dough rise.

GETTING A RISE

1 Mix two tablespoons of dried yeast with one teaspoon of sugar and a cup of warm (not hot) water. Leave this mixture until it starts to foam.

WHY IT WORKS

The yeast mold contains chemicals called enzymes. These react with the sugar to release bubbles of carbon dioxide. The warm water speeds up this reaction. The bubbles of gas cause the bread to rise and take shape.

BUBBLES OF
CARBON DIOXIDE

BREAD RISES

2 Mix five cups of flour, two tablespoons of sugar, four teaspoons of salt, and two tablespoons of butter in a large bowl. Add the yeast mixture and stir in some warm water to form the dough.

3 Place the dough mixture on a floured board and knead it for a while.

28

4 Rinse, dry, and lightly oil the bowl and place the dough in it. Cover the bowl with plastic wrap and leave it in a warm place for a few hours.

5 During this time, the dough will rise. When it has finished rising, take the dough out of the bowl and knead it again until it becomes firm. Mold the dough into loaf shapes.

6 Place the dough loaves on a lightly oiled baking pan, brush them with salty water, sprinkle with a little flour, cover with plastic wrap, and leave them to rise again.

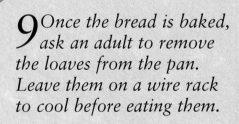

7 Ask an adult to preheat an oven to 430°F (220°C). Remove the dough loaves from the plastic wrap and ask an adult to place them and the baking pan in the oven for 30 to 40 minutes.

8 To check if each loaf is baked, ask an adult to use a damp dishcloth to lift it from the pan and tap the base. If it sounds hollow then the bread is ready. If not, leave it in the oven a little longer.

9 Once the bread is baked, ask an adult to remove the loaves from the pan. Leave them on a wire rack to cool before eating them.

UNLEAVENED BREAD

Try making the bread without adding the yeast mixture. You'll see that the dough does not rise, and you will be left with what is known as unleavened bread.

GLOSSARY

ABSORB (ub-SORB) To take in and hold on to something. *In the project on pages 8–9, you can find out about some substances that water can absorb.*

ACID (A-sud) This is a type of substance that usually has a sour taste. Vinegar and lemon juice are both acids. *Make your own acid-detecting fluid on pages 22–23 and find out which substances are acidic.*

ALKALI (AL-kuh-ly) This is a type of substance that is usually soapy to the touch. Baking powder and milk are both alkalis. *The project on pages 22–23 shows you how to detect alkalis.*

BOILING (BOY-ul-ing) Boiling occurs when the temperature rises and a substance changes its state from a liquid to a gas. *Find out how boiling can separate salt and water in the project on pages 12–13.*

DISSOLVING (dih-ZOLV-ing) Dissolving happens when two substances, such as salt and water, combine completely. When one substance dissolves in another, the result is called a solution. *Look at how some substances can dissolve in water while others can't in the project on pages 8–9.*

FILTER (FIL-tur) To take something unwanted from water. *The project on pages 10–11 shows you how to filter soil from water.*

GLOSSARY

FREEZING (FREEZ-ing) Freezing is when a substance changes its state from a liquid to a solid as the temperature drops. *Look at freezing in action in the project on pages 6–7.*

ORGANIC (or-GA-nik) A reaction is organic when it involves living organisms. *The project on pages 28–29 shows you how living organisms help you to make bread.*

SOLUTION (suh-LOO-shun) This is a fluid in which one substance has dissolved completely in another one. *The project on pages 12–13 shows how to separate a solution of salt and water.*

SURFACE TENSION (SER-fis TEN-chun) The force that holds the surface of a liquid together. *The project on pages 14–15 shows you how to break up the surface tension of water.*

SUSPENSION (suh-SPENT-shun) A suspension is when the particles of one substance float inside another fluid substance. *The project on pages 8–9 shows how fine sand particles float in water in a suspension.*

TEMPERATURE (TEM-pur-cher) How hot or cold something is. *The project on pages 6–7 shows you how water changes at different temperatures.*

INDEX